AF454153

Best Drought Tolerant Grass Plant

For Hot Desert Climates Zone

Bilingual Edition

by

Jannah Firdaus Mediapro

Cyber Sakura Flowers Labs

2022

Jannah Firdaus Mediapro

Publishing

2022

Prolog

Best Drought Tolerant Grass Plant For Hot Desert Climates Zone Bilingual Edition In Englih and Germany Languange.

In many parts of the country or areas with water restrictions, drought-tolerant grasses are recommended for their ability to withstand extended periods without water.

Certain species of grass are better equipped to handle drought because of their native conditions and some grasses are improved cultivars, bred for their drought resistance.

Drought-resistant grasses are one part of a drought-tolerant lawn, along with healthy soil and proper cultural practices.

Drought tolerance is the natural ability of a plant to maintain its growth in arid conditions.

Some plants are naturally adapted, having the ability to handle dry weather on their own, while others have been genetically engineered to be this way.

Below are some excellent low-maintenance grass options to choose from.

In vielen Teilen des Landes oder in Gebieten mit Wassereinschränkungen werden trockenheitstolerante Gräser empfohlen, da sie längere Zeiträume ohne Wasser aushalten können. Bestimmte Grasarten sind aufgrund ihrer ursprünglichen Bedingungen besser für Trockenheit gerüstet, und einige Gräser sind verbesserte Züchtungen, die auf ihre Trockenheitsresistenz hin gezüchtet wurden.

Dürreresistente Gräser sind ein Teil eines trockenheitstoleranten Rasens, zusammen mit einem gesunden Boden und den richtigen Kulturmethoden. Trockentoleranz ist die natürliche Fähigkeit einer Pflanze, ihr Wachstum unter trockenen Bedingungen aufrechtzuerhalten. Einige Pflanzen sind von Natur aus so angepasst, dass sie selbst mit trockenem Wetter zurechtkommen, während andere gentechnisch so verändert wurden, dass sie diese Fähigkeit besitzen. Im Folgenden finden Sie einige ausgezeichnete pflegeleichte Gräser zur Auswahl

1. Bermuda Grass
(Cynodon Dactylon)

Bermuda grass, or bermudagrass, as it is sometimes called, is one of the most popular warm-season grasses you can grow. It performs best in full sun and responds quickly to a bit of water after a period of drought.

Bermudagras oder Bermudagras, wie es manchmal genannt wird, ist eines der beliebtesten Gräser der warmen Jahreszeit. Es gedeiht am besten in voller Sonne und reagiert schnell auf ein wenig Wasser nach einer Dürreperiode.

2. St Augustine Grass
(Stenotaphrum Secundatum)

St Augustine grass is another popular option. This grass has a coarse texture and grows well in dappled shade. It's great for areas that receive only moderate amounts of traffic. St Augustine does not remain green when it is dormant in the winter months. St. Augustine thrives in coastal areas because the weather does not get too cold so the grass stays green.

Augustine-Gras ist eine weitere beliebte Option. Dieses Gras hat eine grobe Struktur und wächst gut in schattigem Gelände. Er eignet sich hervorragend für Flächen, die nur mäßig beansprucht werden. St. Augustine bleibt nicht grün, wenn es in den Wintermonaten in der Ruhephase ist. St. Augustine gedeiht gut in Küstengebieten, da das Wetter dort nicht zu kalt wird und das Gras grün bleibt.

3. Zoysia Grass

(Zoysia Japonica)

Zoysia grass does well in both shade and sun but grows relatively slowly, especially compared to other warm-season grasses like St. Augustine. Once it is established, however, its lush blades will give you a gorgeous turf that tolerates foot traffic with ease.

Zoysia-Gras gedeiht sowohl im Schatten als auch in der Sonne, wächst aber relativ langsam, vor allem im Vergleich zu anderen Gräsern der warmen Jahreszeit wie St. Augustine. Sobald es sich jedoch etabliert hat, erhalten Sie mit seinen üppigen Halmen einen herrlichen Rasen, der problemlos begangen werden kann.

4. Centipede Grass

(Eremochloa Ophiuroides)

Centipede grass produces a unique apple — or lime-green color. Despite being slow-growing, it is attractive and low-maintenance. It tolerates acidic soil and partial shade. It has phenomenal shade tolerance and is a good choice for planting beneath pine trees.

Das Tausendfüßlergras hat eine einzigartige apfel- oder lindgrüne Farbe. Obwohl es langsam wächst, ist es attraktiv und pflegeleicht. Es verträgt sauren Boden und Halbschatten. Es hat eine phänomenale Schattentoleranz und ist eine gute Wahl für die Bepflanzung unter Kiefern.

5. Bahiagrass (Paspalum Notatum)

Bahiagrass is one more warm-season grass for you to consider. It is an all-purpose grass with an excellent ability to tolerate foot traffic. Not only is it one of the best in terms of drought tolerance, making it a good choice for homeowners with active water restrictions, but it also has phenomenal insect and disease resistance.

Bahiagras ist ein weiteres Gras der warmen Jahreszeit, das Sie in Betracht ziehen sollten. Es ist ein Allzweckgras mit einer ausgezeichneten Fähigkeit, Fußverkehr zu tolerieren. Es ist nicht nur eines der besten Gräser in Bezug auf die Trockenheitstoleranz, was es zu einer guten Wahl für Hausbesitzer mit aktiven Wasserbeschränkungen macht, sondern es hat auch eine phänomenale Resistenz gegen Insekten und Krankheiten.

Best Drought Tolerant Grass Plant for Hot Desert Climates Zone Bilingual Edition

6. Wheatgrass
(Pascopyrum Smithii)

Wheatgrass is a cool-season grass that tends to be somewhat coarse-looking. It requires minimal water or fertilizer and is an all-purpose grass that is easy to start from seed.

Weizengras ist ein kühljähriges Gras, das eher grob aussieht. Es benötigt nur wenig Wasser oder Dünger und ist ein universell einsetzbares Gras, das sich leicht aus Samen ziehen lässt.

7. Buffalo Grass
(Bouteloua Dactyloides)

As you might expect from the name, buffalo grass is native to the Midwest and is popular for its infrequent need for mowing and overall hardiness.

It grows lush and thick, requiring only a quarter of an inch per water a week during the summer for optimal growth (though it can survive on less).

This cool-season grass starts relatively slowly, so plan on buying it in plugs rather than starting from seed.

Wie der Name schon vermuten lässt, ist Büffelgras im Mittleren Westen beheimatet und wegen seines geringen Mähbedarfs und seiner allgemeinen Widerstandsfähigkeit beliebt.

Es wächst üppig und dicht und benötigt für ein optimales Wachstum im Sommer nur einen halben Zentimeter Wasser pro Woche (es kann aber auch mit weniger auskommen).

Dieses kühle Gras beginnt relativ langsam zu wachsen, daher sollten Sie es eher als Steckling kaufen, als es aus Samen zu ziehen.

8. Sheep Fescue Grass
(Festuca Ovina)

Another cool-season grass to consider is sheep fescue. An alternative lawn grass type, it grows in clumps, offering a natural look and requiring minimal water.

It requires little mowing and fertilizing but does have a bumpy surface that may not be the best for backyard recreation.

Ein weiteres Gras für die kühle Jahreszeit ist der Schafschwingel. Dieser alternative Rasentyp wächst in Büscheln, bietet ein natürliches Aussehen und benötigt nur wenig Wasser.

Er muss nur wenig gemäht und gedüngt werden, hat aber eine holprige Oberfläche, die für die Freizeitgestaltung im Garten nicht unbedingt geeignet ist.

9. Tall Fescue Grass

(Festuca Arundinacea)

The last of the cool-season grasses on our list is tall fescue. This is one of the most popular grasses of them all with each plant growing from a single seed. Seed heavily and mow often — and you will love the carpet effect produced by this grass type.

Der letzte der Gräser der kühlen Jahreszeit auf unserer Liste ist der Rohrschwingel. Dies ist eines der beliebtesten Gräser überhaupt, denn jede Pflanze wächst aus einem einzigen Samen. Wenn Sie viel säen und oft mähen, werden Sie den Teppicheffekt lieben, den diese Grasart erzeugt.

10. Blue Grama Grass
(Bouteloua Gracilis)

Blue grama is a warm-season perennial grass native to the prairies of North America. It's a hardy species that is highly drought resistant and an excellent choice for homeowners in the western United States looking to use native species in their lawns.

Blue Grama ist ein mehrjähriges, warmes Gras, das in den Prärien Nordamerikas beheimatet ist. Es ist eine robuste Art, die sehr trockenheitsresistent ist und eine ausgezeichnete Wahl für Hausbesitzer im Westen der USA ist, die einheimische Arten in ihrem Rasen verwenden möchten.

11. Pampas Grass
(Cortaderia Selloana)

Native to South America, Pampas Grass (Cortaderia sclloana) is considered one of the most spectacular ornamental grasses. Extremely eye-catching, it features magnificent, dense tussocks of arching, narrow green leaves with margins like razors, that are gracefully topped from late summer to midwinter by huge, silky, silvery white, cream or dusty pink plumes.

Welcomed in any garden style, it is versatile and can be used as a specimen plant, in groups or en masse to create dramatic backdrops, view barriers or wind screens.

It is ideal in the landscape where it provides texture, color and contrast, great in meadows. Tolerates almost all soil conditions, drought, once established, as well as salty and dry winds.

Das in Südamerika beheimatete Pampasgras (Cortaderia selloana) gilt als eines der spektakulärsten Ziergräser. Es ist ein echter Blickfang mit seinen prächtigen, dichten Büscheln aus bogenförmigen, schmalen, grünen Blättern mit messerscharfen Rändern, die vom Spätsommer bis zum Winter von riesigen, seidigen, silbrig-weißen, cremefarbenen oder staubig-rosa Federn gekrönt werden.

Sie ist in jedem Gartenstil willkommen und kann als Solitärpflanze, in Gruppen oder in Massen verwendet werden, um dramatische Kulissen, Sichtbarrieren oder Windschutz zu schaffen. Sie ist ideal für die Landschaft, wo sie Textur, Farbe und Kontrast bietet und sich hervorragend für Wiesen eignet. Sie verträgt fast alle Bodenverhältnisse, Trockenheit sowie salzige und trockene Winde, sobald sie sich etabliert hat.

12. Oriental Fountain Grass (Pennisetum Orientale)

Low-growing, compact and incredibly floriferous, Pennisetum orientale (Oriental Fountain Grass) is a clump-forming tufted perennial grass and one of the most striking hardy fountain grasses. Blooming over an exceptionally long season, its fluffy flower spikes rise high above the foliage, arching upward or outward, and cascading down the deep green to gray-green foliage that rustles in the wind.

They sway in the breeze and literally glow when lit from behind in early morning or late afternoon sun. Over time, those pink panicles mature to a light brown. The foliage turns straw-colored in the fall and generally remains attractive, well into the winter. Performs best in full sun, in average, medium moisture, well-drained soils.

Tolerates part shade, but flowering will not be as colorful. Sandy loams with good drainage are preferred. Drought and deer resistant!

Das niedrig wachsende, kompakte und unglaublich blühfreudige Pennisetum orientale (Orientalisches Brunnengras) ist ein büschelbildendes, mehrjähriges Gras und eines der auffälligsten winterharten Brunnengräser.

Die über eine außergewöhnlich lange Saison blühenden, flauschigen Blütenähren ragen hoch über dem Laub auf, wölben sich nach oben oder außen und fallen kaskadenartig über das tiefgrüne bis graugrüne Laub, das im Wind raschelt.

Sie wiegen sich im Wind und leuchten förmlich, wenn sie in der frühen Morgen- oder späten Nachmittagssonne von hinten angestrahlt werden. Mit der Zeit verfärben sich die rosafarbenen Rispen in ein helles Braun.

Das Laub färbt sich im Herbst strohfarben und bleibt im Allgemeinen bis in den Winter hinein attraktiv. Die Pflanze gedeiht am besten in voller Sonne auf mittelschweren, gut durchlässigen Böden. Sie verträgt auch Halbschatten, blüht dann aber nicht so farbenfroh. Sandige Lehmböden mit guter Drainage werden bevorzugt. Widerstandsfähig gegen Trockenheit und Rehe!

References

Ashraf, M. (January 2010). "Inducing drought tolerance in plants: recent advances". Biotechnology Advances.

"Biotechnology for the Development of Drought Tolerant Crops - Pocket K

Hu, Honghong; Xiong, Lizhong (2014-04-29). "Genetic Engineering and Breeding of Drought-Resistant Crops". Annual Review of Plant Biology.

NAKASHIMA, Kazuo; SUENAGA, Kazuhiro (2017). "Toward the Genetic Improvement of Drought Tolerance in Crops". Japan Agricultural Research Quarterly.

Varshney, Rajeev K; Tuberosa, Roberto; Tardieu, Francois (2018-06-08). "Progress in understanding drought tolerance: from alleles to cropping systems". Journal of Experimental Botany.

Grass Phylogeny Working Group II (2012). "New grass phylogeny resolves deep evolutionary relationships and discovers C4 origins". New Phytologist.

Clayton, W.D.; Renvoise, S.A. (1986). Genera Graminum: Grasses of the world. London: Royal Botanic Garden, Kew. ISBN 9781900347754.

Gibson, David J. (2009). Grasses and Grassland Ecology. Oxford University Press.

Author Bio

"And it is He who sends down rain from the sky, and We produce thereby the growth of all things. We produce from it greenery from which We produce grains arranged in layers.

And from the palm trees – of its emerging fruit are clusters hanging low. And [We produce] gardens of grapevines and olives and pomegranates, similar yet varied.

Look at [each of] its fruit when it yields and [at] its ripening.

Indeed in that are signs for a people who believe."

(From The Holy Quran)